IMAGES
of America
NORMAN'S NAVY YEARS
1942–1959

The Moore-Lindsay Historical House Museum is operated by the Cleveland County Historical Society. (Courtesy of the Cleveland County Historical Society.)

ON THE COVER: In September 1943, the third war-bond campaign was underway. The Cleveland County quota was $1,128,000. Residents were urged to buy bonds at several activities: a war-bond auction, a victory pie supper held at community buildings, and this "buy a bond for a Jeep ride" promotion. The push was on; bond promoters told residents that their goal had not been reached— they had only collected $726,688. (Courtesy of the Cleveland County Historical Society.)

IMAGES
of America
NORMAN'S NAVY YEARS
1942–1959

Sue Schrems and Vernon Maddux
on behalf of the Cleveland County Historical Society

ARCADIA
PUBLISHING

Published by Arcadia Publishing
Charleston, South Carolina

Printed in the United States of America

Library of Congress Control Number: 2015954831

For all general information, please contact Arcadia Publishing:
Telephone 843-853-2070
Fax 843-853-0044
E-mail sales@arcadiapublishing.com
For customer service and orders:
Toll-Free 1-888-313-2665

Visit us on the Internet at www.arcadiapublishing.com

*This book is dedicated to the men and women who were
stationed at the naval air station and naval air technical training
center in Norman, Oklahoma, from 1943 to 1959.*

CONTENTS

ACKNOWLEDGMENTS

If you stop by Ozzie's Restaurant at Max Westheimer Airport in Norman, you will see colorful photographs on the walls of Stearman biplanes used to train Navy pilots at the naval air station that once occupied Max Westheimer Airfield. We often wondered what it was like to hear the planes flying overhead as pilots practiced landings, takeoffs, and circle eights. There are only a few buildings left that mark the Navy's presence in Norman from 1942 until final deactivation of the bases in 1959.

We would like to thank the following people who helped us to collect the photographs and historical materials to compile this history: Bob Rice, for sharing his collection of photographs with us, and the Norman Chamber of Commerce, for donating photographs and historical newspapers that helped us to gain an understanding of life in Norman during the 1940s. We would also like to thank T. Jack Foster Jr. of San Mateo, California, for donating his scrapbook from 1942 to 1943, which highlighted the work of his father, T. Jack Foster Sr.; the Norman Chamber of Commerce, and Savoie Lottinville, director of the University of Oklahoma Press, in bringing the Navy to Norman. We want to thank the membership of the Cleveland County Historical Society, who donated artifacts, volunteer time, and financial support to the historical society and the Moore-Lindsay Historical House Museum. We would also like to thank the City of Norman Town Council for their continued financial support of the Moore-Lindsay Historical House Museum.

Key to Photo Credits:

Cleveland County Historical Society (CCHS)
Western History Collection, University of Oklahoma (WHC)
Bob Rice Photographs (BR)

INTRODUCTION

The September 3, 1939, *Norman Transcript* headline read, "Britain, France Set Ultimatum Deadline," and the September 7 headline read, "Nazis Control One-Third Of Poland." The headlines and the stories that followed told Americans that it was only a matter of time before their country would be drawn into another world war to help European allies fight the aggression of Hitler's Nazi Germany. The question was, would America be up to the challenge after a period of isolationism and a downsized military after World War I. As early as 1938, it was increasingly evident that the United States would need to prepare and strengthen its ability to help European allies fight back Germany's sweep across Europe. T. Jack Foster of the Norman Chamber of Commerce believed that the university and the city could play a role in helping America prepare for war by establishing training programs for Navy and Army reserves, and perhaps establishing an aeronautical program at the university. Within two years of Foster's suggestions, the Japanese attacked Pearl Harbor on December 7, 1941. Within a matter of days after the attack, the United States was at war in the Pacific and in Europe.

The war increased the demand for industrial materials to manufacture warplanes, munitions, supplies, tanks, and ships. This opened factories that were closed during the Great Depression that plagued the country in the 1930s. New manufacturing and an expanded economy put men and women to work. The need for a larger military and training facilities drove the economy. Military officials from all branches of the armed services scouted for locations across the country to build training centers for men and women who joined the military to fight the war. There was a lot of competition from communities like Norman to secure military contracts. University leadership saw it as a way to gain students to replace those who left the university to join the armed forces. The Norman Chamber of Commerce saw military contracts as a way to increase payrolls and commerce in Norman. Several men played key roles in convincing the military to establish training centers in Norman: T. Jack Foster, member of the chamber of commerce and former mayor of Norman; Joseph Brandt, president of the University of Oklahoma; Savoie Lottinville, director of the University of Oklahoma Press; and Neil Johnson, Norman businessman and member of the chamber of commerce. These men fought hard to convince several branches of the military to establish programs in Norman.

The Department of the Navy's decision to establish a naval air station (NAS) in Norman centered on the 288-acre Max Westheimer Airfield, owned by the university. The Navy leased the airfield from the university and eventually purchased additional acreage to equal 1,500 acres. Government contracts went out to architectural firm Leonard H. Baily in Oklahoma City, and construction contracts to Harmon-Cowan and Norton-Tankersley Construction in Shawnee; men were hired, and payrolls expanded. Work started almost immediately on the naval air station, just north of the city. Shortly after construction was underway, the Department of the Navy announced that it was also building a naval air technical training center (NATTC) in Norman. The first site considered was 12 miles south of the city and university. The university and chamber of commerce, with T. Jack Foster's contacts in Washington, convinced the Navy to locate the technical training center just southeast of the university. Eventually, the training center accommodated 20,000 people.

The university, the city, and the chamber of commerce could not imagine the changes that would take place with an increase of so many people in such a short span of time. The town of 11,000 people in 1940 doubled overnight. The town council had to consider how the city's infrastructure was going to handle the extra people; there would be a need for more firefighting facilities and law enforcement. The council considered a bond election to help pay for an expansion in water, sanitation, fire, and police services. The Navy, however, with the help of the 1930s New Deal Works Projects Administration, built more water wells than Norman or the university had operating. The Navy also paved roads that bordered both bases and provided shore patrol to oversee Navy personnel behavior off-base. The Navy even lobbied for improved transportation, which was lacking even with the interurban line running adjacent to the air station.

From May 1942 until June 1945, Norman was a Navy town. But just as quickly as the Navy came to town, by 1944 it looked like the Navy would leave, and that would be an economic blow to the city. The city and the chamber of commerce lobbied the Navy to consider keeping its training facilities in Norman. The Navy was undecided and leaning toward keeping the training centers open. But George Lynn Cross, president of the University of Oklahoma in 1945, wanted the university to take over the land and buildings occupied by the Navy. At the naval air station, the university wanted the control tower, the aircraft hangars, the runways, and the classroom and barracks to be used by university students and academic departments. NATTC Norman, or South Base, adjacent to the university, would provide much-needed housing for veterans and their families, who would be returning to school on the GI Bill. Cross's political pull was stronger than the city's or the chamber of commerce, and the Navy moved its training center to Memphis, Tennessee.

The Navy, however, was not ready to relinquish all to the university. It signed a revocable lease with the university for the land and buildings at the technical training center. When the Korean War started on June 25, 1950, the Navy reactivated many of its training centers across the nation, and by 1952, NATTC Norman was reactivated. President Cross relinquished the former technical training center and moved academic departments back to the main campus. The Navy was back in town. From 1952 to 1959, the Navy continued to train aircraft technicians at the base in Norman. In 1959, it decommissioned the base with little fanfare. The university and the City of Norman were at odds with one another over who would receive naval properties; both petitioned the General Services Administration (GSA) for land, facilities, and water wells. A compromise was reached. The city obtained use of water wells at the air station, acquired the Navy drill field, which became Reaves Park, and the recreation building and swimming pool at South Base. The university was given title to the technical training center and the naval air base.

By the time the Navy left Norman in 1959, the city had an improved infrastructure and a more solid economy than in 1940. The chamber of commerce continued to maintain the city's economy by partnering with the university in a research park at the former naval air station. Today there are very few buildings left at either of the Navy bases. The partnership between the US Navy, the University of Oklahoma, and the City of Norman was a great success. It helped win the war, pulled Central Oklahoma out of the Great Depression, and made the university a great institution.

One

THE NAVAL AIR STATION

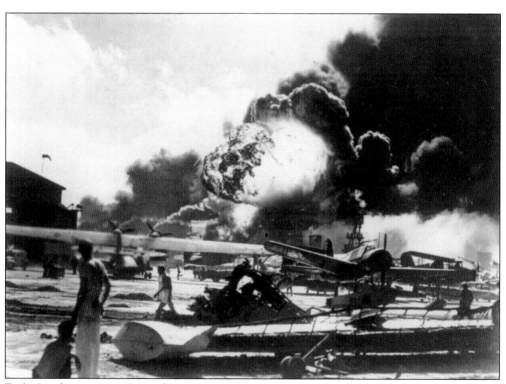

Early Sunday morning December 7, 1941, a Japanese strike force of 31 ships, six of which were aircraft carriers, and 25 submarines streamed toward the American fleet anchored at Pearl Harbor, Hawaii Territory. Minutes before striking Pearl Harbor, Japanese bombers hit Kaneohe Naval Air Station, pictured here. (CCHS.)

The tragedy of Pearl Harbor, with the loss of 2,300 sailors and marines, was brought home to Oklahoma when it was learned that the Japanese strike force bombed and capsized the battleship USS *Oklahoma* (pictured). The *Oklahoma* was commissioned in 1916 and served in World War I protecting allied convoys in the Atlantic. The ship was refitted between 1927 and 1929. In 1936, the Navy assigned the *Oklahoma* for service in the Pacific. (CCHS.)

In July 1942, the Navy started to salvage the *Oklahoma*. It took eight months to right the hull. The *Oklahoma* was decommissioned on June 16, 1943. In 1946, a group of Oklahomans found the *Oklahoma*'s gold punch bowl in storage at Puget Sound Naval Shipyard. The bowl is decorated with the state seal and images of the 1889 Land Run and wheat stalks. Today, the bowl is on display at the Oklahoma Historical Society. (CCHS.)

STREET SCENE. NORMAN,OKLA.

At the time of the Pearl Harbor attack, Norman had approximately 11,000 residents. Norman citizens struggled with the economic uncertainties of the Great Depression of the 1930s and the inevitability of another war on the horizon. The need for defense programs after Pearl Harbor and the US declaration of war provided the impetus for economic recovery in Norman and Oklahoma. (CCHS.)

As early as 1938, T. Jack Foster, vice president of the Norman Chamber of Commerce, thought that the United States was destined to become involved in a new European war. He believed that the University of Oklahoma should be part of the defense program for the war, especially the university's aeronautical-engineering department. He also emphasized that an adequate airport would be needed for civilian training. (CCHS.)

Foster considered Cimarron Field in Yukon an ideal location for an airport, but there was not enough land available for an airfield. Officials at the University of Oklahoma, who realized the importance of a flight-training program for the security of the nation, took the initiative and developed a program with a financial gift from the Max Westheimer estate to purchase land for a flying field in 1941. (CCHS.)

University students indicated an interest in flying as early as 1927, when they organized Tau Omega, an aviation fraternity promoting aeronautical studies. The fraternity established the first flying school in Norman. The Walter Neustadt gift of $10,500 from the Max Westheimer estate allowed the university to purchase 160 acres north of Norman. The City of Norman purchased 128 additional acres and leased them to the university. (CCHS.)

When the United States entered the war in Europe and the Pacific, the university immediately saw the effects of war by the numbers of students who left to join the armed services. In the spring of 1941, enrollment at the university was 5,431; in the spring of 1942, only 4,609 students were enrolled, a reduction of 822 students. The reduction affected tuition, class size, and faculty assignments. (CCHS.)

In January 1942, the university and chamber of commerce presented their arguments to the Department of the Navy for development of a defense preflight school at the University of Oklahoma. The Navy officials in Washington, DC, were not convinced. The university and the chamber realized they would have to personally take their request directly to Washington. (BR.)

Bringing armed-service training programs to the university would fill the student gap. It would also benefit the City of Norman. It was expected that the influx of new students and programs would have a positive economic impact on the town and the region. The question was how big of a training program, and did the city have the infrastructure to handle a large increase in population. (CCHS.)

In February 1942, T. Jack Foster and Neil Johnson of the chamber of commerce, along with Savoie Lottinville of the University of Oklahoma Press, traveled to Washington to lobby for a preflight training school in Norman. The Navy's objections concerned air space, length of runways, desirable weather, and available housing. (CCHS.)

Along with T. Jack Foster, Savoie Lottinville (center) was influential in bringing the Navy to Norman. As reported in George Lynn Cross's book *The University of Oklahoma and World War II*, Lottinville's involvement started with his train trip to New York City in January 1942. Navy commander Kenneth B. Salisbury shared a sleeping compartment with Lottinville. During their conversation, Salisbury asked if the university had a flying program. Lottinville mentioned the Max Westheimer–funded flying field. Salisbury indicated that the Navy was looking for locations to establish training bases across the country. Salisbury invited Lottinville to visit the Department of the Navy in Washington, DC. Joseph Brandt, president of the University of Oklahoma, encouraged Lottinville to visit with the Navy about selecting Norman for a naval training base. Lottinville's account omits the efforts of T. Jack Foster and Neil Johnson. Lottinville was in Washington, but he and Johnson left when the possibility for a base looked bleak. Foster, however, stayed two months and wrote 18 different proposals. In the end, Foster convinced the Navy Site Selection Committee to visit Norman. (CCHS.)

University officials envisioned training courses on campus; they never imagined a separate base. The Navy picked a site for the training school that was 12 miles southeast of Norman. Officials from the university and the city urged the Navy representatives in Norman to choose a site closer to Norman and the university. Ultimately, T. Jack Foster's contacts in Washington convinced the Navy to build near the university. Eventually, there were two military training courses at the

university: the V-5 program, with specialized training courses, and V-12, with more general courses leading to a baccalaureate degree. Many Norman citizens were alarmed when they heard that it was possible that 20,000 men would be stationed in town. Parents were concerned that so many young men might be a threat to the morals of their young daughters. (CCHS.)

The Navy Site Selection Committee agreed that Max Westheimer Airfield had possibilities for a Navy flight school, but it would need more than 288 acres. In all, the Navy acquired 1,500 acres. President Brandt went to Washington in late March 1942 to talk with the Navy about the airbase. When he returned, he announced that the Navy would also establish a school to train mechanics. (CCHS.)

Flood Avenue and the Santa Fe Railroad and interurban tracks bordered Max Westheimer Airfield and the naval air station on the east. Robinson Street bordered the station on the south. The Navy focused the base entrance to face Flood Avenue, with the airfield runways at the west end of the base. The recreation hall and swimming pool were dominant features facing Flood Avenue. (CCHS.)

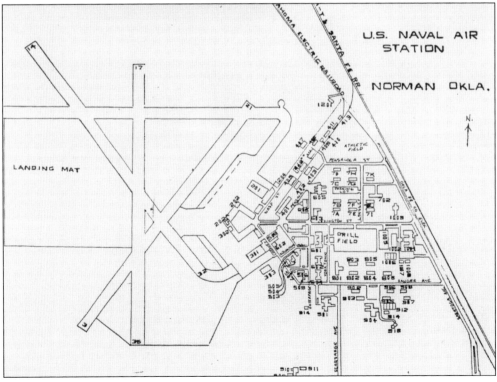

When the Navy finished the bases north and south of town, the sleepy town of 11,000 citizens increased by over 20,000 Navy personnel. The city launched a campaign to encourage citizens to clean their yards and offer fair rent to servicemen and women. The city council also took action to make sure the water and sewage systems could handle the increase in population. (BR.)

The city council meeting of April 1942 dealt with what the increase in population would mean to the city's resources. A city resolution passed to assess fire, police, water, sewage, drainage, housing, roads, and recreational facilities. The city called for a bond election to provide the financial resources to meet the requirements of the Navy. (BR.)

On April 13, 1942, Lt. Comdr. R.B. Meade of the Navy Bureau of Yards and Docks arrived in Norman as officer in charge of construction of the naval air station (NAS). Plans called for steel beams and trusses, but the hangar, or "air garage," pictured here was constructed out of wood, saving valuable steel for aircraft and ship construction. The hangar was completed in August 1942. (CCHS.)

Leonard H. Bailey of Oklahoma City designed the buildings at the air station, seen here. Construction contracts were awarded to Harmon-Cowan and Norton-Tankersley Company of Shawnee and Oklahoma City. The original construction cost was estimated at $5,082,292.70. New contracts and station projects increased construction cost to $9,774,018.98. The station eventually consisted of 91 buildings. (CCHS.)

Procurement of construction materials for the base was a major undertaking. Norman lumberyards and local suppliers' stock was quickly depleted; contractors had to search across the country for construction materials. In June, shipments started to arrive by trucks and on the Santa Fe Railroad. The pace increased throughout the summer of 1942. (CCHS.)

In August 1942, some 50 to 60 trainloads of supplies arrived each day. The spur of the Santa Fe Railroad into NAS Norman was filled with supply cars, as were spurs to Oklahoma City and Purcell. There were 52 firms contracted for naval construction in Cleveland County. One of the biggest jobs was laying the asphalt runways. Other subcontractors drilled wells, installed electrical lines, graded, and performed general earthwork. (CCHS.)

The air station's administration building housed the offices of the base commander. The first commander of NAS Norman, in March 1942, was J.W. Williams Jr. During World War I, Williams had been a flight instructor at the naval air station in San Diego. Between the two world wars, he was an orange grower in California. With the expansion of the naval cadet program in 1935, the Navy called Williams back into service. (CCHS.)

Capt. William N. Updegraff took command of the air station in October 1942. He was commander of NAS Dutch Harbor in Alaska during the Japanese attack in June 1942. For his actions against the Japanese, he received the Navy Cross. After leaving Norman, he married Norman resident Froma Johnson in Hawaii in August 1946. Froma's brother, Neil Johnson, was instrumental in bringing the Navy to Norman. (CCHS.)

On July 31, 1942, a flag-raising ceremony officially opened the naval air station. Until the base was ready to be occupied, the ship's company was stationed at the Norman Armory. They were moved to the air station in September 1942. The base was originally commissioned as a naval reserve aviation base. It kept that designation until January 1943, when it was rechristened a naval air station. (CCHS.)

While flight training was underway, construction continued at NAS Norman. By mid-October 1942, medical resources were moved from the administration building to the new dispensary, and in November the recreation hall (pictured) was completed. It housed the ship's store, station theater, library, bowling alley, public relations, and welfare department. (CCHS.)

The recreation hall served as the hub of social life at the station. Navy personnel enjoyed pre-released movies, big bands, dancing, and lectures by flying aces fresh from the Pacific theater. Church services were held in the recreation hall before the chapel was finished. In front of the recreation hall was the bus stop, where the station bus carried sailors to the main gate, their barracks, or town. (BR.)

The bachelor officers' quarters (BOQ) was actually five barracks clustered around a central building. The barracks consisted of bedrooms and sitting rooms or suites for officers. The central building was a community center and clubhouse with a dining room. The central building was for officers and their guests and contained a lounge and game rooms for billiards, Ping-Pong tables, and card tables. (CCHS.)

Beginning in July 1942, officers from the Marine Corps Reserve Aviation Unit from Squantum, Massachusetts, and from the naval air station in San Diego, California, arrived at NAS Norman. Major Ralston from Massachusetts became the superintendent of aviation training and base ordnance officer. The Marine Corps Aviation Unit at NAS Norman was formally established on September 12, 1942. (CCHS.)

Many of the first Navy recruits came from Oklahoma. According to one officer in *History of United States Air Station Norman, Oklahoma*, "When the call went out for recruits, hundreds of men in the vicinity climbed down off tractors, quit jobs in offices, shops, and factories to answer the call." By August 1942, some 1,400 Oklahomans began eight weeks of boot camp at the unfinished base and graduated the first week of October 1942. (CCHS.)

In September 1942, the first class of Navy cadets arrived at NAS Norman from Chapel Hill, North Carolina. At the time, only 50 percent of the base was complete. The cadets are pictured here lined up for inspection by Lt. C.W. Mitchell Jr., naval security officer. Every 90 days, 800 cadets graduated from the flight-training school. (CCHS.)

Starting in June 1942, the academic qualification for applicants to flight school was a high school diploma. An applicant accepted into the program had to attend one of 17 preparatory schools set up at various college campuses. The curriculum included elementary mathematics and physics, aircraft engines, theory of flight, navigation, aerology, communications, and rigorous physical training. The first class of Navy cadets to arrive at NAS Norman is seen here. (CCHS.)

The first plane to touch down at the air station landed on August 10, 1942. The primary trainer was the Boeing N2S Stearman, pictured here. The plane was designed before World War II and modeled after rugged biplanes that crossed the continent carrying airmail. The Stearman was usually the first airplane a student flew as a cadet. Its construction was durable, with a steel tube fuselage with wood and fabric wings. (BR.)

At the naval air station, cadets received their primary flying instruction. There were four basic models of student trainers at NAS Norman, but only three were in primary use: the Stearman, the Spartan NPS, and the Timm N2T-1. Ground school subjects included navigation, aerology, recognition, gunnery, code, and blinker. In recognition, cadets learned characteristics of a plane and had to recognize them in a .04-second exposure. (CCHS.)

The naval air station had 17 auxiliary flying fields for pilots to practice takeoffs and landings. One of the fields was at Goldsby, a small town just south of the South Canadian River. This flying field was the most advanced of the outlying fields in that it had longer runways. The smaller fields were used for touch-and-go training. (CCHS.)

After a three-month course, the cadets moved on to advanced training at bases such as Corpus Christi, Texas. Training there was in an advanced aircraft and included instrument flight and advanced navigation. Training was not without risks. Cadet William Phillips remembered a cadet from his hometown being killed while practicing a figure eight. (CCHS.)

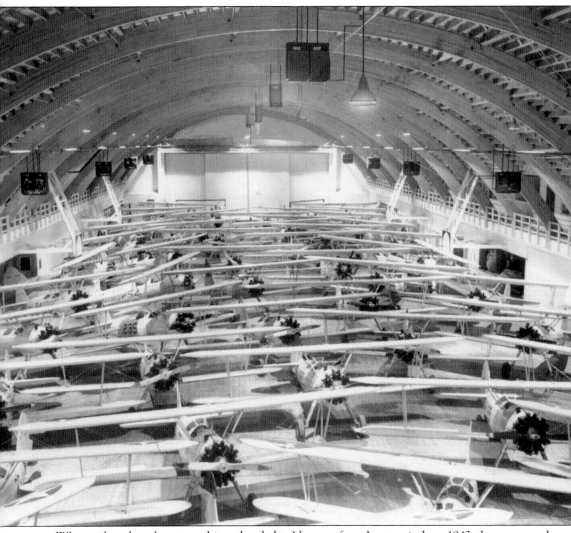

When sailors thought a tornado was headed to Norman from Lawton in June 1943, they crammed 53 planes in the hangar in 15 minutes. The tornado did not materialize. The hangar, pictured here, was also used as the air station's gymnasium. There was enough floor space for six regulation basketball courts. The hangar was used for athletic events, cadet classes, entertainment, and several major shows and big band events. (CCHS.)

For those who lost their lives while training at NAS Norman, a section of the administration building was reserved as Memorial Hall. Between June 1942 and August 1945, twenty-two officers, fifteen aviation cadets, and nine enlisted men died in aircraft accidents at NAS Norman. Sailors created the memorial out of noncritical materials. (CCHS.)

Navy construction crews finished the swimming pool in July 1943, one of the last buildings completed. The pool cost $160,000, and had a depth of four feet, six inches to nine feet. The pool had a sliding roof for indoor or outdoor use; it was positioned behind the base's recreation hall. (CCHS.)

Flight-training courses included water-safety instruction in the indoor pool. The water course included underwater swimming, swimming with flight clothing, lifesaving, high diving, and life-raft procedure. The primary use of the pool was to prepare trainees for complete water adaptability for routine duties and any emergency at sea. (CCHS.)

After basic skills training in swimming and rescue at sea, the men entered into the rugged water-training phase. The pools served other purposes as well. Station personnel had access to the pools for swimming lessons and recreational swimming. According to the Navy, the outdoor pool was the largest in the American Southwest. (CCHS.)

The gunnery department's outdoor range was a manmade hill at the western edge of NAS Norman called Mount Williams in honor of the first base commander, J.W. Williams Jr. This hill was used for rifle and machine-gun practice and other ordnance. Mount Williams continued to be a Norman landmark until the early 21st century, when it was leveled to make room for new business development at I-35 and Robinson Street. (BR.)

Under Mount Williams was a bunker. The ordnance department had the latest model weapons in 1942. The boot school students received training with the 20 Eddystone rifles, which were used on security watch. The beginning of the gunnery program was in the armory in downtown Norman in a 14-by-14-foot room. Later, the program was moved from the old armory to hangar No. 1 at the air station. (CCHS.)

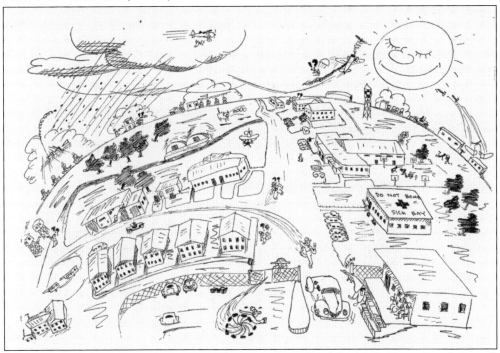

Purcell, Oklahoma, was the site of a naval air gunnery school (NAGS). Even though Purcell was the official designation, the school was located five miles east of Lexington. The school was primarily for antiaircraft gunnery training. The ranges for training included trap and skeet range, three moving-target ranges, a primary range, a sighting range, a turret-performance range, a .22-caliber range, a pistol range, and a short range. (CCHS.)

The naval air gunnery school was not under the direct command of NAS or NATTC Norman. The school was under the command of the regional headquarters of naval technical training command in Chicago, Illinois. Purcell and Lexington suffered the same housing shortage that Norman experienced. Citizens were encouraged to make their spare rooms into rentals rooms for personnel. (CCHS.)

One sailor at NAGS Purcell related that guns occupied so much of his time that he "dreamed about guns." Another sailor commented, "We started our shootin' on the shotgun range and after six weeks of this, single fifties and thirties, from the ground, aboard trucks, and on platforms, with a little pistol packin' thrown in we thought we were about as good gunners as Uncle Sam could find." (BR.)

When the Navy's site-selection committee visited Norman, before they decided to establish a naval base in the city in 1942, they were concerned about the weather conditions and if the climate was suitable for flight training. University and city officials commented that most days, the weather was dry and clear. Not much was said about the spring storms that brought damaging winds and tornadoes, as seen here in 1945. (BR.)

In September 1945, Central Oklahoma was battered with 93-mile-per-hour winds. Officials at the base called the storm freakish. In 1945, having weather instruments to predict storms was still in the future. The windstorm lasted 40 minutes and was followed by heavy rains. The storm's damage indicates there may have been tornadoes in addition to high-velocity winds. (BR.)

The storm tore planes from their moorings and carried them down the runways. One witness related that the planes were stacked in twisted piles three to four high and that "wings were snapped off, tailpieces mangled and fuselages smashed flat." In all, the storm damaged 175 planes. (BR.)

The damage was so widespread that by the next morning, officials were lamenting the cost of training in Stearman planes. But most of the planes were put back into commission; only 25 were a complete loss as flight trainers. Salvage crews were busy for months cannibalizing parts from the wrecked planes. (BR.)

Stearman planes took the brunt of the damage, but heavy service planes were also damaged by the wind. The windstorm presented an opportunity for the galley to operate under battle conditions, as they served coffee and sandwiches to the crew working to clean up the storm's destruction. (BR.)

On April 30, 1949, a storm southwest of Norman produced several tornadoes. The 179th Infantry Division at the base, with 65 national guardsmen, were on the artillery range when the tornado struck the naval air station. The men only had a few seconds' warning—most took cover in the bunker under Mount Williams, seen here in 1993. In all, five men were killed and 48 injured. (CCHS.)

Two

THE NAVAL AIR TECHNICAL TRAINING CENTER

The naval air technical training center (NATTC) was located southeast of the University of Oklahoma and was locally known as South Base. In April 1942, Joseph Brandt, in his capacity as president of the University of Oklahoma, informed the university's board of regents that the prospects looked good for Norman to obtain a naval training center as well as the naval air station. (CCHS.)

President Brandt sent William H. Carson, dean of the mechanical petroleum engineering school, to Washington, DC, to work out the details with the Navy to establish the training center in Norman. Dean Carson ultimately received a commission in the Navy and became liaison officer between the Navy and the university. The east gate to NATTC Norman was just west of Highway 77, as seen in this aerial view. (CCSH.)

The Navy planned to acquire over 1,300 acres southeast of the university—south of Lindsey and west of Highway 77. What was once thought by the university administrators to be a small program to train men on the university campus blossomed into the establishment of a small city to train and house 20,000 men and women. (CCHS.)

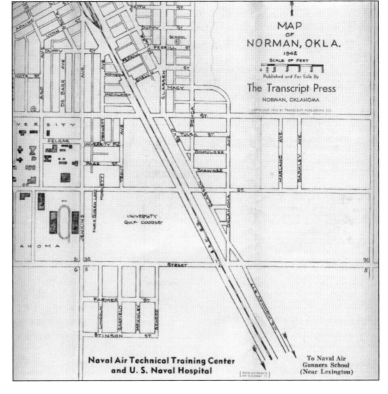

42

In May 1942, construction crews operating under Navy contracts were notified that the construction of an aviation service school would be added to previous naval contracts for construction of the naval air base. The contractors started to work on the land purchased by the Navy, which was a cornfield southeast of the University of Oklahoma campus. (CCHS.)

In December 1942, the Navy commissioned the air technical training command, later renamed the naval air technical training center in February 1943. The command at NATTC Norman was responsible to the chief of naval air technical training in Chicago, who in turn reported to the Bureau of Aeronautics in Washington. (CCHS.)

Lt. Comdr. Norman S. Gallison wasted little time assigning sailors to classes in October 1942, even though instruction lacked the basic equipment. Until the equipment arrived, teaching-aid charts and drawings were used to supplement hands-on training. The University of Oklahoma band played at the commissioning of the base in December 1942. The Navy gradually completed construction of South Base buildings during the early part of 1943. (CCHS.)

While the base was under construction, the base newspaper, *The Bull Horn*, began publication on November 12, 1942. The newspaper informed sailors about important matters on the base and provided information on entertainment. The station held a contest among the sailors for the name of the newspaper. YN3 R.E. Rampy won first prize, $7 in canteen checks. (CCHS.)

The Norman Chamber of Commerce, invested in creating goodwill between the city and the Navy, presented Lieutenant Commander Gallison with a silver punchbowl service. A punchbowl was a traditional gift often given by Navy crew to the commanding officer at the launching of a new ship. (CCHS.)

On December 1, 1942, Lieutenant Commander Gallison turned over command of the training center to Virgil C. Griffin (pictured) who was the former chief of staff under Adm. William Frederick Halsey Jr. Halsey was commander of the Third Fleet in the Pacific in 1944. Captain Griffin was distinguished for being the first to pilot an airplane, a Vought VE-7, off of an aircraft carrier, the USS *Langley*, in 1922. (CCHS.)

The mission at NATTC Norman was to train men and women as ground crew to maintain naval aircraft. When their three-month coursework was completed, they graduated as naval aircraft technicians with expertise as aviation ordnancemen, aviation machinist's mates, aviation metalsmiths, rubberized equipment repair, and aviation radar operators. Classrooms were connected by outdoor covered walkways. (CCHS).

The newly constructed men's barracks included dormitories, a recreation hall, showers, washrooms, and a mess hall. Although bleak-looking shortly after construction, landscaping was immediately implemented to improve the living areas. Even so, many men thought the buildings were flimsy and complained when windstorms blew dust into their living quarters. The sailor who sent this postcard home marked his room with an X. (CCHS.)

When off duty and not in class, sailors could relax by going to the canteen, which also served as a store and for recreation. The Navy issued a safety warning to watch out for water moccasins in the creek, especially in the summer months. Evidently, the snakes came up a creek from the Canadian River to the south of the base. (CCHS.)

When the Navy hospital was commissioned in 1942, Capt. L.B. Marshall was commanding officer. The hospital was located in the southeast corner of the training base. The hospital grounds included 56 buildings connected with outside covered walkways. Most of the buildings were wards, with the administration building in the front. (WHC.)

A Navy band plays at the dedication of the $3-million naval hospital in 1942. The *Norman Transcript* of January 17 reported that the hospital had the "finest equipment available and a carefully selected personnel of Navy doctors and nurses make the inland U.S. Navy hospital a perfect haven for injured and ill Naval personnel." (WHC.)

The hospital opened with 600 beds; later, more wards were added. The facility was one of the biggest inland hospitals established by the Navy. There were 50 nurses, 50 to 60 medical officers, and 160 corpsmen. Standing in front of the naval hospital are, from left to right, Capt.L.B. Marshall, naval medical officer in command of the hospital, Ens. Doris Anderson, and Ens. Janet Ronk. (WHC.)

The medical facilities provided to Navy personnel also included an outpatient and dental clinic that treated naval dependents. Expectant mothers benefited from the medical services provided at the Hospital—40 to 50 babies were born in the maternity ward each month. Pictured are Ens. James Richard Kinsella, right, with his wife, Jean Alyce Kinsella, seated, and their new newly born son, James Kevin Kinsella. (Courtesy of Mike Kinsella.)

British actress Olivia de Havilland, best known for her role as Melanie in *Gone with the Wind*, was one of many performers who visited service personnel in military hospitals in order to boost the morale of wounded or sick servicemen. By 1944, hospital trains regularly stopped in Norman with causalities from the Pacific theater. (CCHS.)

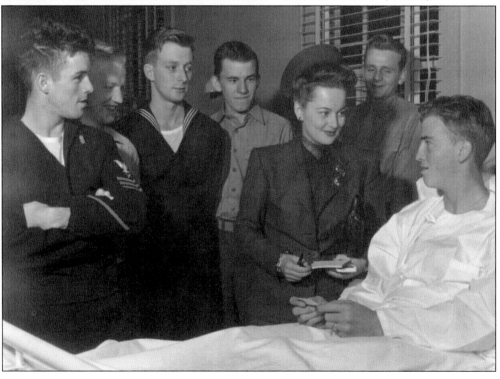

The naval hospital was independent from the technical training center, the air station, and the gunnery school. Together, they were the four Navy installations in Cleveland County. As seen on this map, an overlay of Highway 9 runs through the land once occupied by the hospital. (BR.)

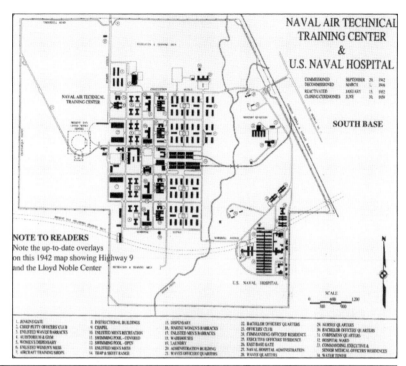

NAVAL AIR TECHNICAL
TRAINING CENTER
&
U.S. NAVAL HOSPITAL

COMMISSIONED	SEPTEMBER	29,	1942
DECOMMISSIONED	MARCH	1,	1946
REACTIVATED	JANUARY	15,	1952
CLOSING CEREMONIES	JUNE	30,	1959

SOUTH BASE

NAVAL AIR TECHNICAL
TRAINING CENTER

NOTE TO READERS
Note the up-to-date overlays on this 1942 map showing Highway 9 and the Lloyd Noble Center

U.S. NAVAL HOSPITAL

SCALE

1. JENKINS GATE
2. CHIEF PETTY OFFICERS CLUB
3. ENLISTED WAVES BARRACKS
4. AUDITORIUM & GYM
5. WOMEN'S DISPENSARY
6. ENLISTED WOMEN'S MESS
7. AIRCRAFT TRAINING SHOPS
8. INSTRUCTIONAL BUILDINGS
9. CHAPEL
10. ENLISTED MEN'S RECREATION
11. SWIMMING POOL - COVERED
12. SWIMMING POOL - OPEN
13. ENLISTED MEN'S MESS
14. TRAP & SKEET RANGE
15. DISPENSARY
16. MARINE WOMEN'S BARRACKS
17. ENLISTED MEN'S BARRACKS
18. WAREHOUSES
19. LAUNDRY
20. ADMINISTRATION BUILDING
21. WAVES OFFICERS QUARTERS
22. BACHELOR OFFICERS QUARTERS
23. OFFICERS CLUB
24. COMMANDING OFFICERS RESIDENCE
25. EXECUTIVE OFFICERS RESIDENCE
26. EAST BASE GATE
27. NAVAL HOSPITAL ADMINISTRATION
28. WAVES QUARTERS
29. NURSES QUARTERS
30. BACHELOR OFFICERS QUARTERS
31. CORPSMENS QUARTERS
32. HOSPITAL WARD
33. COMMANDING, EXECUTIVE &
    SENIOR MEDICAL OFFICERS RESIDENCES
34. WATER TOWER

Civilian ministers, priests, and rabbis held services at the technical training center's chapel. The clergymen, who were commissioned as chaplains in the Navy, had the recommendation of their respective congregations. A chaplain's duties could also include advising and assisting individuals in spiritual as well as domestic, economical, and educational matters. There were typically five chaplains at NATTC Norman. The senior chaplain was also executive secretary of the Oklahoma Auxiliary Navy Relief Society. (CCHS.)

The chapel bell rang the call to worship at NATTC Norman. According to a *Norman Transcript* article, "The bell was brought across the continent with the first wave of Sooners." When presented to the base, an inscription read, "When the slogan of a conquering army is freedom for all mankind, this bell will turn to gold and its tone will become clarion clear." (CCHS.)

The technical training center in Norman was the largest center for naval technical training in the nation. Its mission was to keep the Navy's planes flying no matter the age or the damage. The physical plant, equipment, and number of instructors was equivalent in size to major universities such as Harvard or Yale, according to a Navy manual. (CCHS.)

Pictured here are, from left to right, three unidentified soldiers, Richard Kolonko, Fred Boblanc, and Anthony Avila. Even though NATTC Norman was the Navy's largest naval training center, some sailors were not impressed with the facilities. Eldon Price thought that the two-story frame buildings were stark and the surrounding environs just a red-dirt prairie. The upside was that Oklahoma City was just a trolley ride away. (CCHS.)

Anthony Mennuto, a student at the training center in 1945, wrote his sweetheart Stacie of his excitement at actually getting in the cockpit of the plane he was working on: "We really get some thrilling times once in a while. Last week we started up our first ship (airplane) up to this time we just worked on all the different systems of an airplane but that was really tops." (CCHS.)

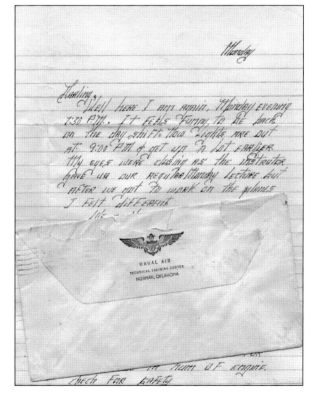

To the aviation metalsmith students, working on the planes was an important job. Anthony Mennuto wrote of his training: "Our purpose in starting the airplanes is to see if the plane is in good enough shape for the pilot to take off. We sit in the cockpit and take notes on the different instrument readings, listen for the neatness in hum of engine, check for safety." (CCHS.)

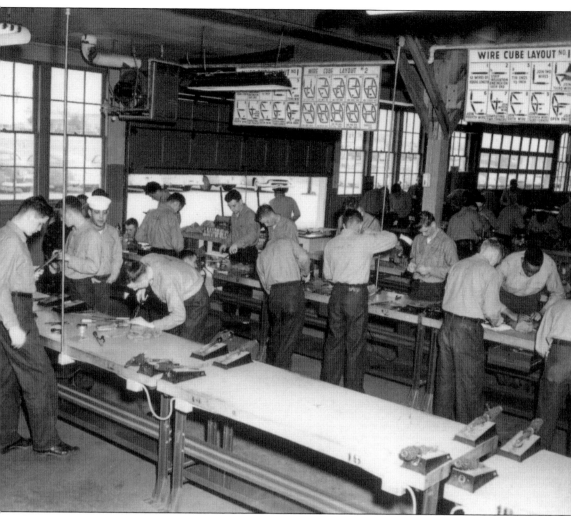

Divisions in the training department, which included aviation ordnancemen, aviation machinist's mates, aviation metalsmiths, rubberized equipment repair, and aviation radar operations, also included a military and physical division. The goal was to develop rugged self-sufficient fighting men. The men went through courses in recognition, signaling, basic ordnance, sighting, and shotgun. (CCHS.)

OPERATIONS PHASE

AVIATION MACHINIST'S MATE
HAND BOOK

RESTRICTED

The aviation machinist's mate school at NATTC Norman was the only one of its kind in the Navy. The students were trained in the manual skill and knowledge necessary to maintain, service, and troubleshoot naval aircraft. The school started with beginning courses but also had courses for sailors stationed at the training center who needed refresher courses. (CCHS.)

In the aviation ordnance school, the goal was to take men straight from boot camp and train them to be of immediate usefulness to the fleet. The 18-week course covered the whole field of aircraft armament. The men started with study of small arms and advanced to machine guns and aircraft cannons. (CCHS.)

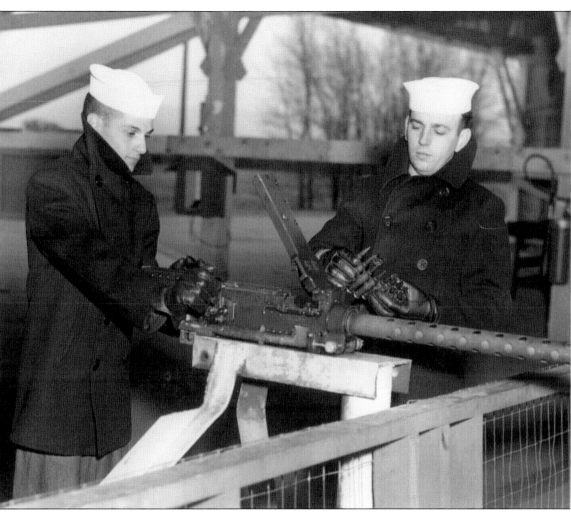

NATTC Norman's primary gunnery program was designed for aviation mechanical mates and aviation ordnance students who qualified for advanced training. Skeet shooting was a particular skill that was relevant in the training program. According to the *Norman Transcript*, the purpose "of the shotgun work is to teach the future gunner the importance of 'lead,' that is shooting at a spot where the enemy plane will be when the bullet reaches its target." (CCHS.)

The Navy opened the training center's pool in May 1943. It was one of three pools planned to provide training and recreation. The pool was 165 feet long by 75 feet wide and could accommodate 100 people. Students took courses to train for routine duties, rugged water, and emergencies at sea. (BR.)

When the United States declared war after Pearl Harbor, both men and women sought ways to help the war effort. It became apparent to the armed services that in order to free up men from desk and shore jobs to fight in Europe and the Pacific, women would have to be trained to take over some of the men's responsibilities. Recruitment posters influenced women to join the armed services. (CCHS.)

Women who joined the Navy's Women Accepted for Volunteer Emergency Services (WAVES) saw it as patriotic to enlist in the Navy. Joan Angle from Detroit, Michigan, stated, "We want to get this war won—quickly." Many women found that joining the armed services was a way of gaining an education and independence before settling into marriage and family life. (CCHS.)

61

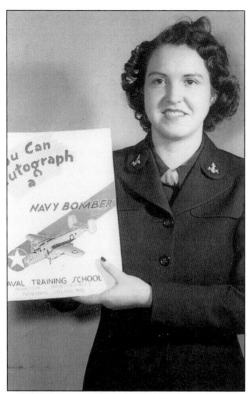

The Navy established the WAVES on July 30, 1942, as a division of the Navy Reserves. Mildred McAfee was the first director. In 1942–1943, some 27,000 women joined the WAVES. Requirements were tough and designed to eliminate those who might not represent the Navy in a good light. Later the requirements were relaxed, and misfits had to be weeded out. (BR)

WAVES arrived in Norman on January 29, 1943. The first detachment came from boot camp at Cedar Falls, Iowa. The officer in charge was Lt. (jg) Myrtle O. Poultney, who prior to the Navy pioneered girls' recreational activities in Hawaii. Poultney also headed several physical-educational programs in California. (WHC.)

After the WAVES first arrived at NATTC Norman, the base newspaper ran several articles on the proper "attitude" between men and women on the base. WAVES were instructed not to let sailors put their arms around them in a movie. Perhaps even better, WAVES were advised to sit in a reserved section of the theater. Male sailors also were warned that they needed to salute a superior WAVES officer. (WHC.)

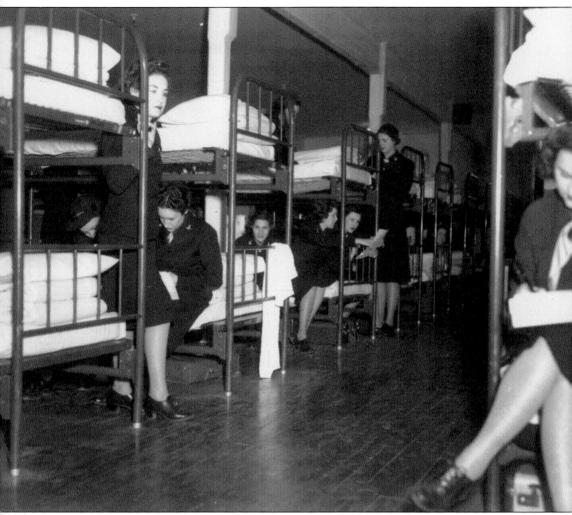

The WAVES barracks were located on the northwest corner of South Base adjacent to the north gate on Jenkins Avenue at Constitution Street. Beyond their barracks, many areas of the base were restricted to them, especially enlisted men's barracks after 1800 hours. The men's barracks were located between Dewy and Porter Avenues and Constitution Street and Merrimac Avenue. (WHC.)

Lt. (jg) Grace Graham was senior barracks officer at the training center. During her duty, she wrote *Long May She Wave*, her reflections on the women she supervised. Her small booklet was in defense of the general perception society had of women who entered the military. Graham allowed, "Their chief ambitions are to do the best job they can now and to become Mrs. America when the peace comes." (CCHS.)

Long May
She Wave!

Grace Graham also reflected in her booklet on women's adjustment to Navy life. To her, there was innocence among the women but strength and a determination to do the same, if not a better job, than the men. WAVES had access to their own "private ships service" (pictured), and in February 1943, the beauty shop opened, which was also available for enlisted men's dependents. (CCHS.)

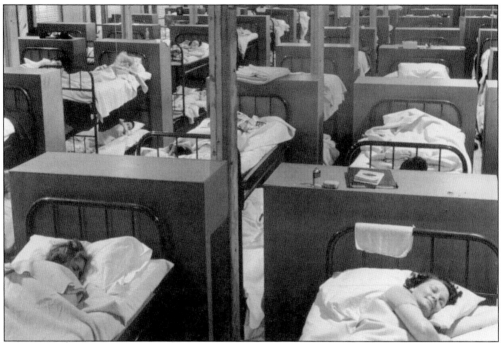

Graham also wrote of women's patriotism: "WAVES are not flag wavers, but there is an undeniable undercurrent of patriotism and idealism in the ranks." WAVES did not receive special treatment because of their gender. According to Graham, if "late for muster or for watch" or the women did "not follow regimental regulations, they received voluntary instruction, which generally consists of cleaning decks, and heads." (CCHC.)

Not long after the WAVES arrived in Norman, a contingent of 50 woman marines arrived from Memphis for duty at South Base. The Marine women had their own barracks and stood out from the WAVES in that they wore different uniforms, a green-and-white seersucker suit. They also had a different salute and training. Marine women were assigned mostly to administrative and clerical work. (BR.)

AD Mary Dodge of Boston, Massachusetts, wrestles with a vise as she learns the skills necessary to qualify for her machinist's mate rating. Prior to joining the WAVES, Dodge was a bank teller in Boston. Not all of the WAVES who reported to duty in Norman were trained to work on aircraft. Some assumed clerical and store positions, freeing men to train for fighting overseas. (CCHS.)

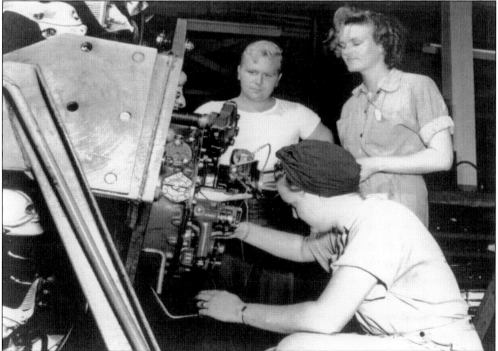

AD1 John Anderson supervises the work of SN1 Inez Waits (foreground) and SN1 Lucille H. Henderson at the naval air station in Jacksonville, Florida. Both women received their basic training at NATTC Norman. A majority of the WAVES in Norman trained as machinist's mates and a small group as metalsmiths. (CCHC.)

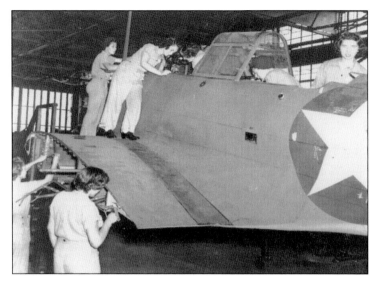

Aviation machinist's mates studied nomenclature, shop mathematics, and blueprint drawings. They also learned to use tools and precision instruments. They learned to check propellers as well as change, repair, and service tires and tubes. Most interestingly, they could start and warm up aircraft engines and learned how to clean and install engines. (CCHS.)

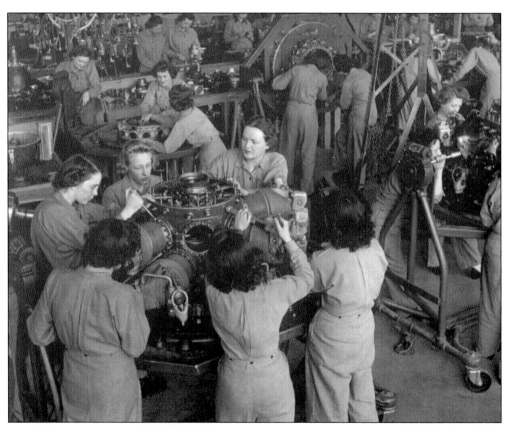

In metalsmith school, WAVES learned some of the basic information as machinist's mates—mathematics and blueprint reading. Their studies also taught them to rivet, weld, and make structural repairs on naval airplanes. When finished with their training, they could assemble and dissemble a plane and adjust and repair the flight controls. (CCHS.)

Along with WAVES who trained to help out in the control tower at the naval sir station, there were six women trained as aerologists at the Lakehurst, New Jersey, aerologist school. Lt. Jane Pryor, assistant officer in aerology, supervised the women. The six WAVES worked eight-hour shifts at the tower. After the war, all had the capability for employment in commercial weather careers. (CCHS.)

Shirley Feldstein Bell (standing third from right) was in the first class of graduating WAVES at the aviation metalsmith school in July 30, 1943. She later wrote of her experience in the Navy: "We went to school with sailors, learning riveting and welding." After her training at NATTC Norman, Bell was stationed at NAS Seattle. "I was the only woman in an Emergency Repair Squadron of 150 men." (CCHS.)

On July 3, 1944, the naval air station planned an open house, in which, according to the *Norman Transcript*, the "former secret facilities will be displayed." Captain Updegraff, station commander, permitted visitors in buildings housing "many synthetic training devices," in particular, the "Link Trainers will be in full operation." For the first time, visitors were able to see the various types of planes in use over Norman and the maintenance facilities. (WHC.)

Citizens of Norman were curious about the naval bases, especially since they benefitted economically from the over $7 million used to construct the base. Also, the $350,000 monthly payroll of base personnel, which equaled expanding Norman by 700 families, stimulated the economy. Citizens approved of the facility and what it meant to the city. (WHC.)

Three

NAVY LIFE OFF BASE

In January 1944, the welfare department at NATTC Norman published a guidebook that provided base personnel with information on Oklahoma. A.L. Simon, a sailor at the training center, illustrated the publication. The booklet was not aimed at providing information about "snooker joints" or tattoo artists, but instead described historical facts about the Dust Bowl, Oklahoma City, Oklahoma City Zoo, and other places of interest. (CCHS.)

Norman established a United Service Organization (USO) council before the Navy arrived in 1942. In August 1942, the council opened the Main Street USO center. In December, another USO opened in the National Guard armory (pictured) located adjacent to the Cleveland County Courthouse and the Santa Fe tracks. The Main Street USO had a different purpose than the armory USO, which was almost a block long and consisted of 60,000 square feet. The armory could accommodate large crowds and group activities. The building had basketball courts, 20 craft hobby rooms, and a snack bar that became one of Norman's popular meeting places. There was a recording studio where service personnel recorded messages to send home. The armory's USO also had physical fitness, ballroom dancing, and free legal service. Norman girls established Girls Service Organization to provide hostesses and dance partners for the servicemen who attended USO dances. The girls also provided shopping services, assisted at the information desk, and gave dance instruction to servicemen. (CCHS.)

Froma Johnson was chairman of the USO Main Street Operating Committee when the facility opened in August 1942. The Main Street USO had a more homey atmosphere for smaller group activities. Programs at the Main Street USO consisted of wives' luncheons, small jewelry classes, progressive bridge parties, swing-dance class, and informal dancing. Today the building houses Mister Robert's Furniture, with a plaque dedicated to the USO. (CCHS.)

The cozy atmosphere at the Main Street USO also provided servicemen and women with a comfortable place to study and write letters. Although the young men and women stationed at the Navy bases in Norman were kept pretty busy, they were homesick for family and friends, and some wrote letters daily in hopes of getting a speedy reply. (CCHS.)

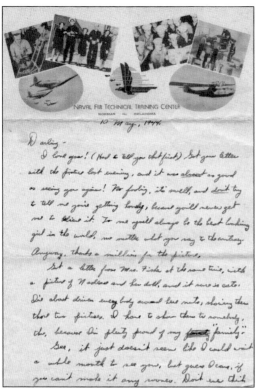

Ira Dean Ogan wrote a letter every day to his sweetheart Joy Thompson during his four months at NATTC Norman. His letters revealed his anguish at not seeing Joy and how he looked forward to her visit to the base. She arrived on a Santa Fe passenger train in May 1944. Ogan left Norman in June 1944 for Seattle. In November 1944, he married Joy in Astoria, Oregon. (CCHS.)

On March 22, 1943, the Great Oak Auditorium officially opened at NATTC Norman with a popular stage show and USO traveling show *Hit the Decks*. The auditorium, building No. 92, was built at the location where a temporary outdoor makeshift stage, old bleachers, and lights that swayed in the breeze served as an entertainment platform for 1,500 people, who sat under the stars and watched the first show. (CCHS.)

The new auditorium (upper left) accommodated 6,000 people. Before the official opening, the first event in the auditorium was a variety show on March 18 featuring the base band, the Gremlins. The variety show included six servicemen and women. The base newspaper related that Gene Serveau of Barracks 70 played his accordion, and nurse Judy Herman, a swing harpist, performed "Tea for Two." (CCHS.)

Most nights, the training center's auditorium offered a variety of entertainment venues that helped to keep up sailors' morale. By day, however, the large auditorium was used for physical-recreation classes. The stands that seated 6,000 were pulled back to reveal basketball courts and room for instruction, as seen in this WAVES dance class. (BR.)

Physical-education programs for WAVES opened soon after they arrived at NATTC Norman. After their 21 weeks of training, the women left Norman more independent and worldly and assured in their strength to do a man's job. Grace Graham, senior barracks officer, commented, "Before I joined the Navy, I was very sensitive; now criticism merely spurs me to correct the situation causing the criticism." (CCHS.)

Shortly after naval cadets arrived in Norman, the University of Oklahoma announced that the Memorial Union would be opened every Sunday afternoon to sailors for card games and movies. The Cotton Club also planned a dance for the cadets; 50 men were invited, and 50 hostesses met them at the door. They danced to the music of Lawrence Welk and his Swing Masters. (CCHS.)

Navy personnel enjoyed first-run movies at the Sooner Theatre. Downtown Norman benefitted from a revived economy when the Navy came to town. Some establishments tried to expand their businesses, especially dance halls and beer parlors. Fred Harris owned a beer parlor in the Fisher Building on Main Street and petitioned the city council to establish a dance hall upstairs. The city had an ordinance prohibiting such ventures. (CCHS.)

In April 1942, the Navy announced that African Americans would be accepted for enlistment in the US Naval Reserves. Norman had an unwritten law against African Americans living in the city; it was expected that they would leave by nightfall. Black sailors who came to town with the Navy, however, successfully challenged Norman's racial segregation customs. (CCHS.)

In all, over 2.4 million African American men and women served in the Navy during World War II. When asked, Lt. Comdr. J.W. Williams stated, "We are all in the war, black and white." There was, however, more segregation than integration at the naval bases in Norman. As seen in this page from the naval air station's yearbook, African American sailors occupied service-related duties at NAS Norman. (CCHS.)

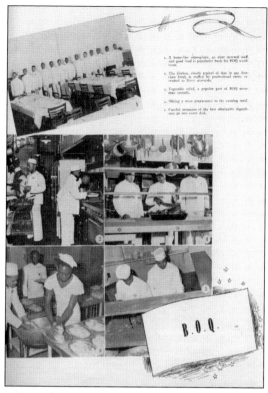

Williams also indicated that African Americans who were stationed in Norman would include a 23-piece all-military band called the Jive Bombers. The name seems to have been a common one for many black bands stationed at naval bases across the country. The group in Norman provided orchestra music for station social functions and band music for drill and inspections. They played at many station dances and auditorium programs. (CCHS.)

The Navy provided dances for African Americans stationed at the naval air station. A newspaper account related, "All Negro military personnel in the Norman area, including both Army and Navy, will be entertained at Naval Air Station, when J. McShann, prominent Negro orchestra leader, brings his band to this community. Women students at Langston University will serve as dancing partners for the men." (CCHS.)

The South Base newspaper, *The Bull Horn*, related news to those at the training center. The May 13, 1943, issue announced "Al Sears' Big All-Star Show in 92 at 2200." *The Bull Horn* also announced upcoming entertainment at the base—three dances and a show each week. All entertainment was planned in the welfare department at each base station. (CCHS.)

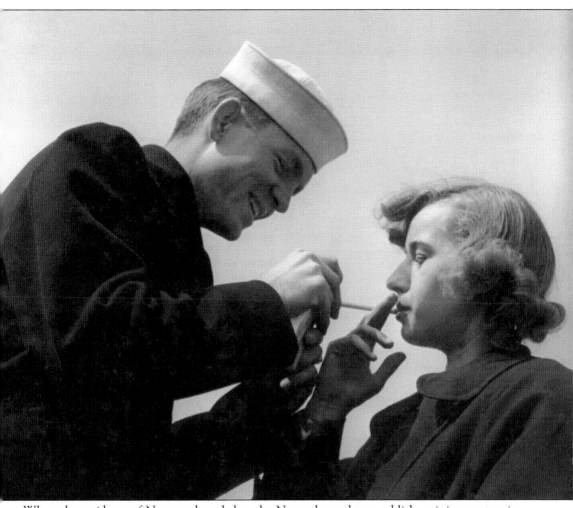

When the residents of Norman heard that the Navy planned to establish training centers in their town, residents had mixed feelings. Jobs would be created, and there would be a boost to the economy, but a significant increase in young men in town was thought by some to be a threat to the town's young women. The Navy put strict guidelines in place regarding the conduct of sailors when "ashore." (CCHS.)

Sailors worked hard and looked forward to liberty, when they could visit downtown Norman or, if they had time, travel to Oklahoma City. While on liberty, men and women conducted themselves according to Navy standards. To keep Navy order, the shore patrol stood as a reminder to stay out of trouble. The interurban station in downtown Norman was one mode of transportation to Oklahoma City. (CCHS.)

Transportation of naval personnel when off base was a major concern for the base commanders. After establishment of the bases, the city did not have sufficient bus transportation, and the Oklahoma Railway Trolley Company (interurban) did not have sufficient cars to transport the thousands of sailors traveling to Oklahoma City. The railway company soon abandoned its express service to convert cars for passenger service. (CCHS.)

The Navy was also concerned about the condition of eating establishments that naval personnel frequented. In December 1942, eight eating and drinking places were banned to sailors. The Navy rated the establishment according to Navy health standards. Eight restaurants were rated "C." These places were out-of-bounds to sailors until improvements were made. Denco Café, established around 1956, was not rated C. (OHS.)

By October 1943, the Navy still rated some Norman establishment unsatisfactory. Three places in particular were Big Bills Bar at 205 South Porter Avenue, Northside Bar at 319 East Main Street, and Southside Bar at 324 East Main Street. The Navy's complaint was that taverns showed "inadequate toilet facilities . . . At the rear of one place with an outside toilet . . .the ground in the area was a cesspool of sewage," according to the *Norman Transcript*. R.M. McCool, city manager, corrected the problem. (OHS.)

Sailors and Norman families gather around Charles V. Martin's train display in downtown Norman on December 19, 1946. Martin was a retired postal worker who had a love for trains; he grew up in Norman not far from the railroad. He started his hobby after he retired by building train cars and scenery in his backyard at 827 North Peters Avenue. (CCHS.)

The train setup in Martin's backyard attracted kids all year round. The display consisted of 1,000 feet of track, scenery, and city buildings. Martin also made steam engines, Pullman, baggage, express, mail, parlor, and dining cars, all of which were made of salvaged materials. The first year that Martin displayed his trains by the depot, the chamber estimated over 13,000 people viewed the exhibit. (CCHS.)

The chamber of commerce sponsored Martin's train exhibit, which was first shown the week before Christmas in 1938 at Edwards Park next to the Norman Depot, now James Garner Plaza. Martin also displayed his trains at the USO for military personnel. This display had 350 feet of track and replicas of city buildings, station houses, bridges, roadways, and advertising billboards. (CCHS.)

Christmas Dinner

NAVAL AIR TECHNICAL TRAINING CENTER
NORMAN, OKLAHOMA

The Navy chapels at the air station and training center provided services for men and women of Catholic, Protestant, and Jewish faiths. The chapels were also used for military weddings. With a town teeming with young men and women, there were many acquaintances and subsequent weddings. One such wedding was held on April 15, 1943. Elaine Forsander of Norman married Lt. Thomas W. Hills, of Lake City, Florida. (CCHS.)

The Christmas dinner menu for 1944 indicates the Navy's effort to give servicemen and women a sense of home and holiday spirit. The dinner started with an ambrosia salad. On the table were radishes, sweet pickles, and hearts of celery. The main course was roasted tom turkey, oyster dressing, giblet gravy, baked spiced sugar-cured ham, mashed cream potatoes, peas, sweet potatoes, cranberry sauce, rolls, and three desserts. (CCHS.)

Sports played an important role in training cadets. Physical-education instructors taught exercise classes and also developed individual performers. The most outstanding athletic program was the football team, the Zoomers, which ranked in the first five service teams in 1944. The baseball team was also outstanding and composed of many former professional and college stars. (CCHS.)

2nd Air Force
SUPERBOMBERS 6
vs.
Norman Naval Air Station
ZOOMERS 13

Owen Field
October 29, 1944
Norman, Okla.

Souvenir
Program
Price 25c

Sports were important to a sailor's training. As one officer related, "A naval aviation cadet must learn the importance of competition. It must be hammered into him until he realizes there is no substitution for winning. That is the way it is in combat. Love of combat, sacrifice for the ideal." (CCHS.)

ZOOMER Staff

Lt. Robert C. Antonides
Assistant Coach

Lt. Edward Jankowski
Backfield Coach

Lt. (jg) Fred Sington
Line Coach

Comdr. Ina Smalling
Executive Officer

Lt. Comdr. John Gregg
Head Coach

Capt. W. N. Updegraff
Commanding Officer

J. H. "Doc" Johnson

Lt. Joe Begala
Assistant Coach

Lt. John J. Economos
Assistant Line Coach

The Zoomers had a qualified staff and many well trained college and professional players. In 1943, the team had 11 experienced players, such as Ens. Steve Andrejko (Androko) of the Washington Redskins. South Base did not have a football team but provided men to play with the North Base Zoomers. The team played regular scheduled games and participated in tournaments. (CCHS.)

Horsemen from the April 1943 89er's Day parade pose in front of the new courthouse, which was built by the Works Progress Administration in 1938. Neil Johnson participated in the ride. Johnson was the grandson of Montford T. Johnson, a Chickasaw rancher, who lived in the territory before white settlement. Johnson was also one of the men who lobbied for a naval training center in Norman. (CCHS.)

Selling US war bonds to finance the war became a community responsibility in towns across America in 1943. Neil Johnson, chairman of the Norman Chamber of Commerce, organized the annual 89er's Day celebration in April 1943 to coincide with Norman's war-bond drive. Here, Johnson speaks to those gathered to celebrate 89er's Day about buying war bonds. (CCHS.)

In September 1943, the third war-bond campaign was underway. The Cleveland County quota was $1,128,000. Residents were urged to buy bonds at several activities: a war-bond auction, a victory pie supper held at community buildings, and this "buy a bond for a jeep ride" promotion. The push was on; bond promoters told residents that their goal had not been reached—they had only collected $726,688. (CCHS.)

The Navy promoted bond sales at both bases and lent jeeps for the campaign in downtown Norman. If a person bought war bonds they would qualify for a jeep ride around town. A newspaper article commented on this photograph: "Comely high school lasses, football players, grade school youngsters waited in line with their war bond purchase for a jeep ride." (CCHS.)

When the Navy announced the establishment of training centers in Norman, the city and chamber of commerce scrambled to find housing for the increased population. In 1942, the chamber created the Housing Authority to promote modernization and repairs on vacant houses and apartments and to establish fair rent policies. New housing developments provided needed housing, as seen here on North Base Avenue. (CCHS.)

The city understood that 15 percent of the base officers would be married and reside in Norman, totaling at least 350 families who would need off-base housing. Unable to find housing, many construction workers and their families who moved to Norman to work on the bases, lived in trailers and tents because of a lack of rentals. (CCHS.)

The city did a survey and found 90 trailers and 13 tent homes in and around the town. There were two trailer camps at the north and west edges of the city, a camp at Boyd Street and Highway 77, one on Porter Avenue, and a large camp at Daws Street and Santa Fe near the city park (pictured). In total, the camps housed over 400 people. (CCHS.)

NAVAL AIR TECHNICAL TRAINING CENTER ★ NORMAN, OKLAHOMA

A Navy promotional booklet published around 1943 titled "We Keep 'Em Flying" sums up the work of the men and women who trained at the naval air technical training center. The booklet presents an overview of the training of those who called the base home during their three-month stay in Norman. In describing base activities, the booklet relates, "This Navy activity is a modern and self contained City." (CCHS.)

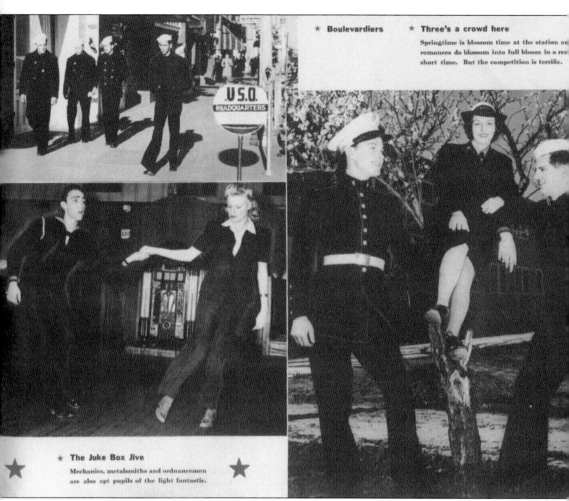

★ **Boulevardiers**   ★ **Three's a crowd here**

Springtime is blossom time at the station and young
romances do blossom into full bloom in a remarkably
short time. But the competition is terrific.

★ **The Juke Box Jive**

Mechanics, metalsmiths and ordnancemen
are also apt pupils of the light fantastic.

"We Keep 'Em Flying" also highlighted the social life of sailors and WAVES when they were off base. The Navy recognized that young men and women in their early 20s would find romance and perhaps long-lasting relationships. As stated on this page, "Springtime is blossom time at the station and young romances do blossom into full bloom in a remarkably short time." (CCHS.)

Four

NAVY BASES AFTER WORLD WAR II

On May 8, 1945, Nazi Germany unconditionally surrendered to the Allies, ending World War II in Europe. On August 9, 1945, Japan surrendered, ending the conflict in the Pacific. The end of hostilities presented the City of Norman and the University of Oklahoma with challenges. T. Jack Foster, representing the chamber of commerce, traveled to Washington to discuss the future of the bases with the Department of the Navy. (CCHS.)

00 - O.U!s "PRE-FAB" Town - Veteran's Ho

University of Oklahoma president George L. Cross eyed the bases and their facilities as necessary for the growth of the university and to help with student housing shortages. Cross believed enrollment could reach 10,000 students; many would be returning veterans and their families. A University of Oklahoma alumnus who owned a lumber-supply business solved a problem by offering surplus materials for prefabricated houses, as shown here. (BR.)

Cross ordered 500 units. There were 200 with two bedrooms and 300 with one bedroom. The cost was $1.25 million. The development became known as Sooner City; it was located south of Jenkins Avenue and west of Asp Avenue. As seen here, construction of Couch Center Towers rises over Sooner City. The university started to phase out the prefabricated village in 1962. The last units were removed in 1966. (BR.)

In January 1946, President Cross learned that the Navy planned to keep naval air station and training facilities in Norman. Cross was adamant in his position that the Navy decommission the two bases. Jack Foster and the chamber of commerce fought to keep the Navy in Norman. Cross solicited the support of Oklahoma governor Robert S. Kerr. The governor endorsed Cross's position in closing the Navy bases. (CCHS.)

President Cross took his case to the press to persuade the citizens of Norman that it was in the best interest of the university that the Navy did not retain the bases. But the citizens of Norman were not convinced. Over 500 people held a rally at the USO armory in downtown Norman protesting the university's move to take over the naval bases. (CCHS.)

With the political pressure from Cross and state officials, the Department of the Navy announced deactivation of their Norman bases on March 1, 1946. All the training programs at NATTC Norman were transferred to Corpus Christi, Texas, or Jacksonville, Florida. In a revocable agreement with the Navy, the university obtained most of the training center and 200 buildings and equipment worth over $25 million. (CCHS.)

The university acquired air station buildings and 1,500 acres of land worth over $4 million. With the acquisition of the naval air station, the University of Oklahoma now had a sufficient airport, with 5,000-foot runways, hangars, and a control tower, all built by the Navy. (CCHS.)

The university lost little time in occupying the former naval air station after base closure in March 1946. The men's recreation hall, center front, became the extension division office. Other University of Oklahoma schools that occupied North Base were engineering, architecture, biological survey, geology education, the library, and freshmen classrooms. (CCHS.)

The Norman Chamber of Commerce and the City of Norman did not win their fight with the university to retain naval training facilities in Norman. But the town did benefit from city improvements. The Navy installed water wells and a sewer system and paved or serviced city roads. But most importantly, the city saw increased employment and business during the war years. (CCHS.)

With the outbreak of war in Korea in June 1950, and the American foreign policy intent on stopping communism, the US military started making plans for training a new generation of soldiers. It was only natural that the Norman Chamber of Commerce would renew their campaign for the naval training bases to return to Norman, which meant more jobs and increased payrolls. (BR.)

The US Army was the first of the armed services to show interest in the military facilities in Norman. In December 1950, the Army believed the former naval air station would be ideal to train airmen. The Navy, however, was still considering reactivation of its bases in Norman and would not sign off on the facilities. (CCHS.)

The US Air Force also tried to obtain training facilities at the former naval air station. But by January 1951, the Navy started showing new interest in the Norman bases. Even though the University of Oklahoma occupied buildings on North and South Base as early as 1945, they understood the agreement they had with the Navy was revocable. (CCHS.)

PLANNING CONFERENCE — Capt. Donald E. Wilcox, right, commanding officer of Naval Air Technical Training Center, listens intently, as Comdr. Leslie Stone, representing the chief of the Naval Air Technical Training Command, initiates the conference session between Navy and O.U. officials here today. Dr. G. L. Cross, O.U. president, center, checks official papers. (Transcript Staff Photo.)

OU's Assistance Asked

Navy Designs Base Reactivation Plans

12/27/51

Initial personnel assigned to the Naval Air Technical Training Center will begin arriving about Jan. 15, Navy officials announced during a conference with O.U. representatives here today.

Comdr. Leslie Stone, representing Rear Adm. William David Johnson, the chief of the Naval Air Technical Training Command, said formal recommissioning will be held Feb. 1, as already announced, but details concerning the ceremony must await further re-activation plans now being formulated.

At this morning's parley, University officials agreed to ready, at cost basis, some of the South Base buildings so the Navy can move as soon as possible.

"We can help out in the emergency but, because of the numerous projects we have under way now, the amount of assistance O. U. can give will, necessarily be limited," Dr. George L. Cross, O. U. president, emphasized.

The meeting was designed primarily to give the Navy delegation, representing the top command for naval air technical training, and public works officers an opportunity to view facilities and ascertain work necessary prior to re-activation.

Stone Heads Session

The NATTC chief called the planning session, inviting Cross to have representative O.U. personnel present.

As leader of the naval group, Stone directed the conference with

Wife Is Okie

Sooner State Nothing New For Base CO

12/27/51

Capt. Donald E. Wilcox, Naval Air Technical Training Center's first commanding officer under re-activation, already has a good introduction to Oklahoma.

His wife is a native of the state — the former Jo Belle Draughon of Marietta.

Wilcox was visiting his wife's family at Marietta while on leave from his current duties as chief of staff for the chief of naval air reserve training, Glenview, Ill., and before flying into Norman this morning for a conference of Navy and O.U. officials.

"I know your state pretty well," Wilcox said, hastily pointing out, however, that he himself is a native of "the wilds of New York."

Wilcox is an ardent fisherman and is looking forward to time off from his Navy duties here for an opportunity to drop a hook into some of the state's popular fishing waters.

The tall, lean captain was wearing civilian attire for his first visit to Norman. (Under present regulations of the military, officers and men are permitted to wear civies when they are not on official duty.)

He made his official landing at Max Westheimer field and was met there by Comdr. Leslie Stone, representing Rear Adm. William David Johnson, chief of the Naval Air Technical Training Command and leader of Navy party here for today's conference.

At the O.U. Union Building, Wilcox was officially welcomed to Norman by Dr. George L. Cross, O.U. president, and other former representatives with whom the Navy representatives made plans for re-activation of the south base, starting Jan. 15.

In December 1951, The Navy announced its plans for reactivation of NATTC in Norman; the formal re-commissioning of the base would be held February 1, 1952. George L. Cross, president of the University of Oklahoma, met with Navy officers to negotiate the university's plan to return base buildings to the Navy. University officials agreed that the buildings would be ready as soon as possible. (CCHS.)

The City of Norman was pleased that the Navy reactivated the bases. They anticipated an annual payroll of $20 million. In 1952, the university had a $5-million payroll and Central State Hospital a $1-million payroll. The city also believed it was better equipped to handle the Navy's needs than in 1942. It had two new water wells and an expanded storm sewer system, both of which were inadequate in 1942. (CCHS.)

Reactivating NATTC Norman also meant ensuring sailors had recreational outlets. In June 1952, Allen R. Webster, director of the USO, visited Norman and stated, "Norman is certainly in need of a USO center—my first priority." The newly appointed Norman USO committee received permission from Brig. Gen. Roy W. Kenny to use part of the downtown armory. Sailors from the training center helped to renovate the armory into a recreational center. (CCHS.)

The Navy indicated that the training courses instituted in 1952 would be more efficient and more streamlined than those established during World War II. The goal was to teach sailors new skills in more "practical factors of aviation, and to familiarize the students with their occupation duties," according to the *Norman Transcript*. The students pictured here are learning how to use firefighting equipment. (CCHS.)

The Navy reactivated the bases in Norman on January 15, 1952, and recommissioned the naval air technical training center on February 1, 1952. The east gate, seen here in 1952, was on Constitution Avenue off Highway 77 and west of the Santa Fe Railroad tracks. (BR.)

A color guard made up of University of Oklahoma cadets in the Naval Reserve Officers Training Corps assisted at the flag-raising ceremony at the reactivation festivities. The naval air station and the naval air technical training center were separate entities during World War II. After reactivation in 1952, both bases were a joint concern under one command. (CCHS.)

Navy officials planned the ceremony even though there was a lack of personnel; many servicemen were not yet aboard the station. The band from the naval air technical training center in Memphis, Tennessee, played at the ceremony. The ceremony took place in front of building No. 4, the administration building, which at that time still housed the University of Oklahoma School of Art. (CCHS.)

Rear Adm. William D. Johnson, chief of the training center in Memphis, stands at the microphone dedicating NATTC Norman on January 15, 1952. Beside him is Capt. Donald E. Wilcox, commanding officer of the base. He stated, "We needed this base, existing facilities could not be further expanded to meet our needs." (CCHS.)

In his remarks citing the need for trained personnel, Rear Adm. William D. Johnson commented: "Our defense services have found it a waste of manpower and materials to buy airplanes unless we can keep them flying—15 to 20 trained technicians are required to keep one airplane in operational duty." (CCHS.)

Dedication of the training center included many of those who fought for the Navy's return to Norman. Besides Rear Admiral Johnson (far right) and Capt. Donald Wilcox (second from left) E.F. Forman (not pictured), president of the chamber of commerce, and John A. Pearson (far left), a retired Army colonel who lived in Norman, welcomed the Navy back to Norman. (CCHS.)

On November 5, 1955, some 4,000 sailors at the training center passed in review for Oklahoma City–Norman high school students and the Navy Buddy Program. Receiving the review, are, from left to right, Maj. Gen. Hal Muldrow, commanding general of the 45th division; Capt. L.W. Parrish, commanding officer of NATTC Norman; Phil Kidd, president of the Norman Navy league, and Dr. George L. Cross, president of the University of Oklahoma. (CCHS.)

The naval air technical training center supported its own drill team. The team consisted of 20 men and two officers. Those who qualified for the team were students undergoing the basic aviation course offered by the Navy. The men practiced on their own time and performed at many special occasions around Oklahoma. The team participated in the November 5, 1955, review for high school students and for Navy Buddy Program. (CCHS.)

In January 1952, there were 10 officers and 25 men on base. One year later, there were 4,700 students, 1,950 enlisted personnel, and 100 officers. By 1954, the aviation fundamental school graduated 16,600 students. In 1954, there was a decline in the number of students, with only 7,000 graduates. By the end of 1955, a total of 47,000 students graduated from NATTC Norman. (CCHS.)

After the armistice ended the Korean War in 1953, naval training continued in Norman until the Department of the Navy officially closed the training center in June 1959. Personnel at the bases were relocated to the naval air station in Memphis. Naval personnel in Norman quickly planned for the move and arranged for equipment to be moved out in vans, trailers, and railroad boxcars. (CCHS.)

James K.P. Wilson, junior officer of the day, fills out papers transferring one of the less than 100 enlisted men from the Norman base. Many of the men did not know what they were going to do when they left the Navy. According to the base historian, "They just wanted out." (CCHS.)

Capt. L.W. Parrish, base commander, and his executive officer, Comdr. Russell Pearson, and Chief James K. Wilson perform their last official duty of lowering the flag on June 30, 1959. The small ceremony closed 17 years of naval activity in Norman and a great working relationship between the Navy and the citizens of Norman. (CCHS.)

There was no pomp or ceremony, and no dignitaries represented the chamber of commerce or the university—just a routine lowering of the colors by Captain Parrish (left) and Chief Wilson (right). There were few sailors looking on as the naval air technical training center moved into history. Norman and the rest of the country enjoyed the prosperity of the postwar economy of the 1950s. (CCHS.)

After the final flag-lowering ceremony, Capt. L.W. Parrish cleaned off his desk. The Navy reassigned Parrish to the naval air station in Memphis. He planned to leave the morning after the final flag-lowering ceremony. Captain Parrish lamented, "It's very trying, especially when you've done a lot of things to improve the base, and now you have to rip them out and ship them to some other base." (CCHS.)

It was a routine transfer for the men, but many felt sad to leave Norman, especially to leave the personal ties they had with the city. Lt. A.J. Furnweger headed the crew of 17 sailors and a few civilians who saw to the arrangements of shipping the last remnants of equipment before their departure at the end of August 1959. (CCHS.)

After Lieutenant Furnweger and his crew left the base at the end of August 1959, only a handful of fire and security guards remained to roam the empty 270 buildings and nearly 1,300 acres where the Navy trained 92,000 men since 1952. As a lonely statement, the bell tower and bulletin board stand in disarray in front of the chapel. (CCHS.)

With the closing of the base, 231 civilian workers lost their jobs. The Navy, however, worked to place these employees in civil-service jobs. Most of the civilian workers were placed in government jobs in the Oklahoma City area. Some had to relocate outside of Norman. The last sailors to leave the base left mops as lone survivors. They were too old and worn out to be transferred. (CCHS).

With the closing of the naval air technical training center in 1959, the City of Norman lost an annual payroll of $6 million. But it was a different era, and the chamber of commerce saw many economic opportunities for the city, a city that had made enormous progress since the end of the Great Depression in the 1930s. (CCHS.)

The first item on the city and the university's agenda after the base closings was dividing spoils. One university objective in lobbying for naval bases in Norman was to eventually acquire naval properties. After the base closings, naval surplus properties were assigned to the General Services Administration (GSA) for disposal, which had to be in the interest of the area and community as a whole. (CCHS.)

The GSA informed the university that it could obtain properties without cost if used for educational needs. As for Norman, the GSA allowed the city to buy properties at 60 percent of assessed value. But properties had to be used by the public at no charge, such as land used for parks. (CCHS.)

The Navy developed its own water and sewage system, both superior to that of Norman. North Base had the best water, with eight wells. The water from the 18 wells on South Base was brackish and infiltrated the university's three wells. To remedy the situation, the Navy extended a water line from North Base to South Base and let the university tap into this line. (CCHS.)

After base closings, Norman's mayor June Benson applied to the GSA for title of the water system, including the university's wells. The mayor figured that if the city controlled the water system the university would have to pay for city water. The city also applied for the sewage-disposal system and the 160-acre Navy drill field. The university applied for the same properties. (BR.)

The GSA suggested that the city and the university reach some kind of compromise before applying for naval properties. The final outcome gave the city the drill field, which became Reaves Park (pictured), the Great Oak Recreation Hall, the pool, and the sewage system. The University of Oklahoma obtained the water system but licensed to the city Navy wells. The city could buy surplus university water at cost. (CCHS.)

The University of Oklahoma acquired the airfield tower built by the Navy in 1942. The university also acquired the four runways, hangars, and cafeteria. Students in aeronautical engineering took up residence in the old Navy barracks. Students also used Navy recreational facilities, including the Olympic-sized swimming pool. (CCHS.)

The university received the deed to 607 acres and 47 buildings of the former training center, including the golf course developed during the Navy's hiatus from Norman from 1946 to 1952. With this deed, the Navy's revocable lease signed after deactivation of the bases in 1946 was void. The government retained the remaining land, to be sold later for industrial development. (CCHS.)

In 1958, Laurence S. Reid, chairman of the University's school of natural gas engineering, believed that Oklahoma should develop an industrial research park in Norman. His thought was to develop land for industrial firms to establish permanent research-and-development facilities. Norman had the ideal requirements for such a new concept in 1958: good weather, natural resources, university connections, and naval air station land. (CCHS.)

Reid viewed the land and facilities at the former air base as an ideal location for a research park in that there was developed land, roads, and sufficient buildings. The chamber of commerce viewed a research park as a vehicle to attract industry that would employ at least 4,500 people. The University of Oklahoma's Merrick Computer Center (pictured) was one of the first to build on the former base. (CCHS.)

The University of Oklahoma also envisioned the old naval air station as the nucleus for a research campus. Soon after the Navy left Norman in 1959, the university developed Sevearingen Research Park at the base, now called north campus. Old Navy building No. 604, the instructional building, housed the severe storms laboratory. (CCHS.)

Among the old barracks and instructional buildings on north campus, there were newly constructed radar domes and instrument towers. Names of streets assigned by the Navy such as Lexington, Saratoga, Pensacola, and Yorktown were changed to reflect the ideals of a new era: Einstein, Newton, and Mendel. Scientific inquiry against a backdrop of military buildings denoted a new time and place. (CCHS.)

The weather center and severe storms lab held a prominent place on north campus. Personnel from the National Weather Service, the Federal Aviation Administration, and the Air Force worked at the center, which became a nucleus for a cooperative unit of eight federal, state, and university entities that employed 200 scientists, technicians, and teaching faculty. Today, the new weather center is on the University of Oklahoma's main campus. (CCHS.)

Today, there are only a few buildings left that indicate that there were once two Navy installations in Norman. Over the years, the facilities have been demolished to make way for new buildings representing a different era. Pictured here is construction of the YMCA at the former naval air station. In the background is one of the only buildings left, an old hangar now used for basketball. (CCHS.)

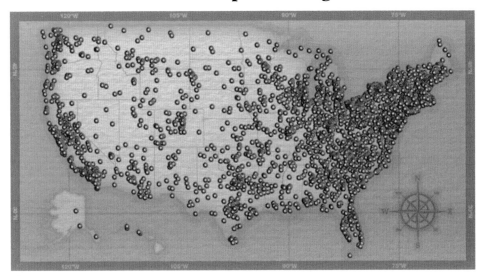